Understanding Bird Flu

Causes, Symptoms, and Transmission

By Oluchi Ike

Preface:

Avian influenza, commonly known as bird flu, has captured global attention due to its potential to impact human health and animal populations. This book aims to demystify bird flu by offering a comprehensive exploration of its causes, symptoms, and modes of transmission. Written for readers with or without a background in medicine or biology, the book blends scientific insights with practical knowledge.

Whether you're a healthcare professional, a student, or someone interested in zoonotic diseases, this guide will equip you with the tools to better understand this pressing health issue.

Table of Contents:

Chapter 1: Introduction to Bird Flu

1.1 What is Bird Flu?

1.2 History of Bird Flu Outbreaks

1.3 Importance of Understanding Bird Flu

1.4 Overview of Avian Influenza Viruses

Chapter 2: Causes and Origins

2.1 What Causes Bird Flu?

2.2 Bird Flu in Wild and Domestic Birds

2.3 Role of Migratory Birds in Virus Spread

2.4 Mutations and Cross-Species Transmission

Chapter 3: Symptoms in Birds

3.1 Identifying Signs of Bird Flu in Birds

3.2 Severity of Symptoms by Strain

3.3 Impacts on Poultry Health and Economy

3.4 Diagnostic Techniques for Avian Influenza

Chapter 4: Symptoms in Humans

4.1 Recognizing Bird Flu Symptoms in Humans

4.2 Complications and Severe Cases

4.3 Differentiating Bird Flu from Seasonal Flu

4.4 Diagnosis and Testing for Human Cases

Chapter 5: Transmission Dynamics

5.1 How Bird Flu Spreads Among Birds

5.2 Human Transmission Pathways

5.3 Zoonotic Risks and Public Health Concerns

5.4 Environmental Factors Affecting Spread

Chapter 6: Prevention and Control in Birds

6.1 Best Practices in Biosecurity

6.2 Vaccination in Poultry

6.3 Surveillance and Rapid Response

6.4 Challenges in Eradicating Bird Flu

Chapter 7: Prevention and Control in Humans

7.1 Personal Protective Measures

7.2 Vaccines and Antiviral Medications

7.3 Public Awareness Campaigns

7.4 Role of Global Health Organizations

Chapter 8: Impact on Global Health and Economy

8.1 Economic Costs of Bird Flu Outbreaks

8.2 Effects on Food Security

8.3 Impact on Wildlife Conservation

8.4 Pandemic Preparedness and Lessons Learned

Chapter 9: Future Outlook

9.1 Research Trends in Avian Influenza

9.2 Genetic Changes and Vaccine Development

9.3 Cross-Sector Collaborations in Disease Management

9.4 Challenges in Predicting Future Outbreaks

Chapter 10: Conclusion and Key Takeaways

10.1 Summary of Key Points

10.2 Importance of Continued Vigilance

10.3 Role of Individuals in Prevention

10.4 Building Resilience Against Emerging Diseases

Chapter 1: Introduction to Bird Flu

1.1 What is Bird Flu?

Bird flu, scientifically known as avian influenza, refers to infections caused by influenza viruses that primarily affect birds. These viruses belong to the Orthomyxoviridae family and are classified as type A influenza viruses. Avian influenza viruses naturally occur among wild aquatic birds and can infect domestic poultry, other bird species, and even mammals, including humans.

Bird flu is categorized based on its pathogenicity, or the ability to cause disease, into two types:

- **Low Pathogenic Avian Influenza (LPAI):** These strains typically cause mild symptoms or no symptoms at all in birds. They can, however, evolve into highly pathogenic forms.

- **Highly Pathogenic Avian Influenza (HPAI):** These strains are highly virulent, leading to severe disease and high mortality rates in birds.

The most well-known bird flu strain, H5N1, has gained global attention due to its ability to infect humans and cause severe respiratory illness. While human-to-human transmission is rare, the potential for such transmission raises concerns about its pandemic potential.

1.2 History of Bird Flu Outbreaks

Bird flu has a long history, with significant outbreaks affecting both avian and human populations. Understanding these outbreaks provides insights into the virus's behavior and its potential impact:

- **1878:** The first recorded instance of avian influenza occurred in Italy, initially termed "fowl plague." At the time, little was known about the virus or its zoonotic potential.

- **1959:** The first HPAI outbreak caused by an H5N1-like virus was documented in Scotland. This event marked a turning point in understanding the highly pathogenic nature of certain avian influenza strains.

- **1997:** A major milestone in bird flu history was the first recorded transmission of H5N1 to humans in Hong Kong. This outbreak involved 18 human cases, with 6 fatalities, and resulted in the culling of 1.5 million chickens to control the spread.

- **2003-2004:** The largest H5N1 outbreak to date began in Southeast Asia, particularly affecting countries like Vietnam and Thailand. This epidemic highlighted the role of migratory birds in the global spread of avian influenza.

- **2013:** The emergence of H7N9, another avian influenza strain in China, resulted in severe human infections. Unlike H5N1, H7N9 did not cause significant disease in birds, complicating detection and control efforts.

- **2021-2022:** Recent outbreaks of HPAI in Europe and North America demonstrated the ongoing threat of bird flu. These events affected millions of birds, causing economic and ecological consequences.

Each outbreak underscores the need for surveillance, rapid response, and international collaboration to mitigate the impact of avian influenza.

1.3 Importance of Understanding Bird Flu

Understanding bird flu is crucial for several reasons:

Public Health Implications

Bird flu poses a significant threat to human health due to its zoonotic nature. While human infections are rare, they often result in severe respiratory illnesses, with high mortality rates.

Studying bird flu helps identify risk factors and develop preventative measures to protect public health.

Agricultural Impact

Avian influenza outbreaks can devastate the poultry industry, leading to significant economic losses. Infected birds are often culled to prevent the spread, disrupting food supply chains and affecting livelihoods. A comprehensive understanding of bird flu enables the development of biosecurity measures to protect poultry populations.

Pandemic Potential

The ability of avian influenza viruses to mutate and reassort genetic material with human influenza viruses raises concerns about their pandemic potential. Such events could result in a highly transmissible and virulent strain, leading to a global health crisis. Monitoring and studying bird flu strains help scientists assess and prepare for these risks.

Ecological and Conservation Concerns

Wild bird populations are natural reservoirs of avian influenza viruses. Outbreaks can affect biodiversity, particularly among endangered bird species. Conservation efforts depend on understanding how these viruses circulate in the wild and impact bird populations.

By prioritizing research and education on bird flu, society can better prepare for and mitigate its impacts on health, economy, and the environment.

1.4 Overview of Avian Influenza Viruses

Avian influenza viruses are diverse and complex, classified based on two surface proteins:

- **Hemagglutinin (H):** This protein helps the virus attach to and enter host cells. There are 16 subtypes (H1-H16) found in birds.

- **Neuraminidase (N):** This protein aids in the release of new virus particles from infected cells. There are 9 subtypes (N1-N9) found in birds.

Combining these subtypes results in various strains, such as H5N1 and H7N9, each with unique characteristics and risks.

Natural Reservoirs

Wild aquatic birds, such as ducks and geese, serve as the primary reservoirs for avian influenza viruses. These birds often carry the virus asymptomatically, enabling its spread across regions through migration. Domestic poultry, such as chickens and turkeys, are highly susceptible to infection, particularly with HPAI strains.

Transmission Mechanisms

Avian influenza viruses spread through direct contact with infected birds, contaminated surfaces, or respiratory droplets. In wild bird populations, the virus can also spread through water contaminated with bird feces. Human infections often occur through close contact with infected poultry or contaminated environments.

Mutations and Reassortment

Avian influenza viruses are known for their ability to mutate and reassort genetic material with other influenza viruses. This adaptability enables them to infect new hosts and, in rare cases, transmit between humans. Understanding these mechanisms is essential for predicting and preventing potential pandemics.

Control and Prevention

Efforts to control avian influenza focus on surveillance, biosecurity, and vaccination. Early detection through monitoring programs enables rapid response to outbreaks. Vaccines are developed for both birds and humans, though their efficacy varies by strain. Biosecurity measures, such as restricting access to poultry farms and disinfecting equipment, play a crucial role in preventing the spread of the virus.

The foundational knowledge in this chapter sets the stage for exploring bird flu in greater detail. Subsequent chapters will delve into the causes, symptoms, and transmission dynamics, providing a comprehensive understanding of this significant health challenge.

Chapter 2: Causes and Origins

2.1 What Causes Bird Flu?

Bird flu, or avian influenza, is caused by influenza A viruses that primarily infect birds but can also infect other species, including humans, under specific circumstances. The causative agents of bird flu belong to the Orthomyxoviridae family, characterized by their RNA-based genetic material. These viruses have a segmented genome, making them highly adaptable through genetic reassortment and mutation.

The primary cause of bird flu infections is exposure to the influenza virus, which occurs through direct contact with infected birds, their secretions, or contaminated environments. Key pathways include:

- **Fecal-Oral Transmission:** Bird flu viruses are often excreted in the feces of infected birds, contaminating water and feed sources.

- **Respiratory Droplets:** Infected birds can spread the virus through sneezing, coughing, or breathing, releasing viral particles into the air.

- **Contaminated Surfaces:** Equipment, clothing, and vehicles exposed to infected birds or their droppings can harbor the virus and facilitate its spread.

Host factors, such as the immune response and genetic susceptibility, also play a critical role in the disease's progression. While wild birds often serve as asymptomatic carriers, domestic birds like chickens and turkeys are highly vulnerable to severe illness.

2.2 Bird Flu in Wild and Domestic Birds

Wild Birds

Wild aquatic birds, particularly waterfowl like ducks, geese, and swans, serve as natural reservoirs for avian influenza viruses. These birds often carry the virus asymptomatically, allowing it to persist in nature without causing widespread mortality in their populations. In wild birds, the virus primarily affects the gastrointestinal tract rather than the respiratory system, facilitating its spread through feces.

Key characteristics of avian influenza in wild birds include:

- **Asymptomatic Carriage:** Wild birds can carry low pathogenic avian influenza (LPAI) strains without showing symptoms, spreading the virus over long distances.

- **Reservoir Role:** They act as a source of infection for domestic birds, particularly when their habitats overlap.

- **Ecological Impact:** While LPAI strains are common, highly pathogenic avian influenza (HPAI) outbreaks can lead to significant mortality in certain wild bird species, disrupting ecosystems.

Domestic Birds

Domestic birds, especially poultry, are highly susceptible to avian influenza. The disease can range from mild symptoms, such as ruffled feathers and decreased egg production, to severe and fatal conditions, especially in HPAI outbreaks. Factors contributing to the high vulnerability of domestic birds include:

- **Close Proximity:** High-density farming practices increase the risk of rapid virus transmission.

- **Genetic Homogeneity:** Limited genetic diversity in domesticated poultry populations can reduce their ability to resist infections.

- **Limited Exposure:** Unlike wild birds, domestic birds often lack prior exposure to low pathogenic strains, leaving them immunologically naive.

HPAI outbreaks in domestic poultry have devastating economic and social impacts, necessitating stringent control measures such as culling and biosecurity enhancements.

2.3 Role of Migratory Birds in Virus Spread

Migratory birds play a pivotal role in the global spread of avian influenza viruses. As natural reservoirs, they contribute to the introduction and dissemination of the virus across continents, often without exhibiting symptoms. This characteristic makes their role in the epidemiology of bird flu both significant and challenging to manage.

Migration Routes and Spread

Migratory birds travel long distances along established flyways, which serve as pathways for virus dissemination. Key flyways include:

- **East Atlantic Flyway:** Connecting Europe, Africa, and parts of Asia.

- **East Asian-Australasian Flyway:** Extending from Siberia and China to Southeast Asia and Australia.

- **Pacific Americas Flyway:** Linking Alaska, Canada, and the Americas.

During migration, birds interact with local populations, potentially introducing new virus strains into previously unaffected areas. Wetlands, estuaries, and other shared habitats become hotspots for virus transmission.

Role in HPAI Spread

Migratory birds are often implicated in the initial spread of HPAI strains to domestic poultry farms. For example, the global spread of the H5N1 strain in the early 2000s was strongly linked to migratory bird movements. This highlights the importance of monitoring migratory patterns and implementing biosecurity measures to reduce the risk of transmission to domestic birds.

Challenges in Management

Managing the role of migratory birds in avian influenza spread poses unique challenges:

- **Monitoring and Surveillance:** Tracking virus prevalence in migratory populations requires international cooperation and advanced technology.

- **Conflict with Conservation Efforts:** Measures to control avian influenza must balance disease prevention with the protection of migratory bird species and their habitats.

- **Unpredictable Patterns:** Changes in climate and habitat availability can alter migratory routes, complicating efforts to predict and mitigate the spread of the virus.

2.4 Mutations and Cross-Species Transmission

One of the most concerning aspects of avian influenza is its ability to mutate and jump species barriers. These processes are driven by the virus's genetic structure and interactions with various hosts, leading to new strains with potentially increased virulence and transmissibility.

Mechanisms of Mutation

Avian influenza viruses mutate through two primary mechanisms:

- **Antigenic Drift:** Small, gradual changes in the virus's genetic material occur during replication. These changes can alter surface proteins, such as hemagglutinin (H) and neuraminidase (N), allowing the virus to evade the host immune response.

- **Antigenic Shift:** A more dramatic process, antigenic shift involves the reassortment of genetic material between different influenza viruses. This can occur when a single host, such as a pig or human, is simultaneously infected with avian and human influenza viruses. The resulting hybrid virus may possess characteristics that enable efficient human-to-human transmission.

Cross-Species Transmission

The ability of avian influenza viruses to infect new host species depends on several factors:

- **Receptor Binding:** The virus must bind to specific receptors on the host's cells. Avian influenza viruses typically bind to alpha-2,3-linked sialic acid receptors, which are abundant in birds. However, mutations can enable binding to alpha-2,6-linked sialic acid receptors, found in the human respiratory tract.

- **Host Susceptibility:** Species-specific factors, such as immune system differences and cellular environments, influence the likelihood of infection.

- **Environmental Factors:** Conditions that promote close contact between species, such as live bird markets and farming practices, increase the risk of cross-species transmission.

Examples of Cross-Species Transmission

Historical events demonstrate the potential for avian influenza viruses to infect new species:

- **H5N1 in Humans:** First reported in 1997, this strain caused severe respiratory illness in humans, with a mortality rate exceeding 50%. Most cases were linked to direct contact with infected poultry.

- **H7N9 in Humans:** Emerging in 2013, this strain primarily affected individuals exposed to live bird markets in China. While human-to-human transmission was rare, the virus caused severe illness and raised pandemic concerns.

Implications for Pandemic Preparedness

The ability of avian influenza viruses to mutate and cross species barriers poses significant challenges for global health:

- **Surveillance:** Continuous monitoring of avian influenza viruses in both wild and domestic birds is essential to detect emerging strains with pandemic potential.

- **Vaccine Development:** Vaccines must be updated regularly to address evolving virus strains. Research into universal influenza vaccines is ongoing to provide broader protection.

- **International Collaboration:** Coordinated efforts are required to share data, resources, and expertise for effective prevention and response strategies.

Understanding the causes and origins of bird flu provides a critical foundation for managing and mitigating its impact. This chapter highlights the complex interplay between environmental, biological, and human factors that drive the emergence and spread of avian influenza, emphasizing the need for a holistic and proactive approach to disease prevention.

Chapter 3: Symptoms in Birds

3.1 Identifying Signs of Bird Flu in Birds

Identifying bird flu symptoms in avian species is crucial for early detection and control of outbreaks. Symptoms of bird flu vary widely depending on the strain of the virus, the species of bird, and individual health conditions. Recognizing these signs can help in prompt diagnosis and prevention of widespread transmission.

Common Symptoms of Bird Flu in Birds:

- **Respiratory Distress:** Birds infected with avian influenza often exhibit difficulty breathing, open-mouth breathing, coughing, and nasal discharge. These symptoms are among the first signs noticed by poultry farmers.

- **Swelling and Cyanosis:** Swelling of the head, combs, wattles, and legs is a hallmark symptom of some highly pathogenic avian influenza (HPAI) strains. Cyanosis, or a bluish discoloration of these areas, is another indicator.

- **Reduced Activity:** Infected birds frequently become lethargic, with reduced movement and unwillingness to eat or drink.

- **Diarrhea:** Watery and greenish diarrhea is a common symptom that can lead to rapid dehydration and weight loss.

- **Neurological Symptoms:** Some birds may show abnormal behaviors, such as circling, head tilting, or paralysis, due to the virus's impact on the nervous system.

- **Decreased Egg Production:** Layers may produce fewer eggs, and the eggshell quality can decline, becoming thin or misshapen.

- **Sudden Death:** In severe cases, particularly with HPAI strains, apparently healthy birds may die suddenly without showing prior symptoms.

Understanding these symptoms and their progression is critical for veterinarians, farmers, and wildlife specialists to implement timely interventions.

3.2 Severity of Symptoms by Strain

The severity of symptoms in birds depends on the pathogenicity of the avian influenza virus. Influenza A viruses are classified into two categories based on their ability to cause disease in poultry:

Low Pathogenic Avian Influenza (LPAI)

LPAI strains generally cause mild or asymptomatic infections. These strains primarily affect the respiratory and gastrointestinal tracts and often go unnoticed in wild birds. However, in domestic poultry, LPAI can lead to:

- Mild respiratory symptoms like sneezing and coughing.

- Decreased feed intake and weight gain.

- Slight reductions in egg production.

While LPAI poses less immediate threat, it has the potential to mutate into highly pathogenic forms under certain conditions.

Highly Pathogenic Avian Influenza (HPAI)

HPAI strains are much more severe and can result in rapid mortality. Notable features include:

- **Acute Onset:** Symptoms often appear suddenly and progress rapidly, leading to death within 48 hours in many cases.

- **High Mortality Rates:** HPAI outbreaks can result in mortality rates nearing 100% in unprotected flocks.

- **Systemic Impact:** The virus spreads beyond the respiratory and gastrointestinal tracts, causing multi-organ failure.

HPAI outbreaks are particularly devastating due to their rapid spread, high mortality, and significant economic repercussions.

3.3 Impacts on Poultry Health and Economy

Avian influenza outbreaks have far-reaching consequences, not only for the health of poultry but also for the livelihoods of farmers and the broader economy. Understanding these impacts is essential for justifying and implementing control measures.

Health Impacts on Poultry

- **Widespread Mortality:** HPAI outbreaks can decimate entire flocks, causing catastrophic losses for poultry producers.

- **Secondary Infections:** Birds weakened by LPAI are more susceptible to secondary bacterial infections, compounding health issues.

- **Decreased Productivity:** Even birds that survive may suffer long-term effects, including reduced growth rates and egg production.

Economic Impacts

- **Loss of Livestock:** The direct loss of birds during outbreaks results in immediate financial losses for poultry farmers.

- **Culling Costs:** Mass culling of infected and at-risk birds is a standard control measure but incurs significant expenses.

- **Market Disruptions:** Trade restrictions and consumer fears can lead to decreased demand for poultry products, further affecting revenue.

- **International Trade Bans:** Countries often impose bans on poultry imports from affected regions, impacting global trade.

Broader Implications

- **Food Security:** In regions where poultry is a primary protein source, outbreaks can threaten food supply chains.

- **Public Health Concerns:** Although rare, the potential for zoonotic transmission heightens public health risks and necessitates additional resources for monitoring and control.

Investing in prevention, surveillance, and rapid response systems can mitigate these impacts, ensuring both poultry health and economic stability.

3.4 Diagnostic Techniques for Avian Influenza

Accurate and timely diagnosis of bird flu is critical for controlling its spread and minimizing its impact. Diagnostic methods vary in complexity, speed, and application, from field-level tests to advanced laboratory techniques.

Clinical Observation

While observing symptoms can provide initial clues, clinical signs alone are insufficient for a definitive diagnosis. Many respiratory diseases share similar symptoms, making laboratory confirmation essential.

Field Diagnostics

- **Rapid Antigen Tests:** These portable tests detect viral antigens in swabs from the bird's respiratory or cloacal tracts. While fast and easy to use, they may lack sensitivity and specificity.

- **Point-of-Care PCR:** Some field kits include portable polymerase chain reaction (PCR) devices, offering higher accuracy than antigen tests but requiring more technical expertise.

Laboratory Diagnostics

- **Virus Isolation:** This gold standard involves inoculating embryonated chicken eggs or cell cultures with samples from suspected cases. The process confirms the presence of the virus but is time-intensive.

- **Molecular Techniques:**
  - **RT-PCR:** Reverse transcription polymerase chain reaction is the most widely used technique for detecting and characterizing avian influenza viruses. It identifies viral RNA with high sensitivity and specificity.

- **Genome Sequencing:** Sequencing provides detailed information about the virus's genetic makeup, helping track mutations and identify new strains.

Serological Tests

- **Hemagglutination Inhibition (HI) Test:** This test detects antibodies against avian influenza viruses in blood samples. It is commonly used for surveillance and post-vaccination monitoring.

- **ELISA:** Enzyme-linked immunosorbent assay is a sensitive and quantitative method for detecting antibodies or viral antigens.

Surveillance Programs

Active and passive surveillance programs are crucial for early detection of avian influenza outbreaks. These programs involve:

- **Sampling:** Regular testing of wild and domestic bird populations.

- **Data Analysis:** Monitoring trends and identifying potential hotspots.

- **International Collaboration:** Sharing data across borders to track and manage global disease spread.

Challenges in Diagnosis

- **Resource Limitations:** Many regions lack access to advanced diagnostic tools and laboratories.

- **Rapid Mutation:** The virus's ability to mutate complicates efforts to develop universal diagnostic methods.

- **Wild Bird Populations:** Monitoring wild birds is logistically challenging but essential for understanding the virus's ecology.

Accurate diagnosis is the cornerstone of effective avian influenza management. By combining clinical observation, field diagnostics, and advanced laboratory techniques, veterinarians and researchers can detect and control outbreaks more efficiently.

Chapter 4: Symptoms in Humans

4.1 Recognizing Bird Flu Symptoms in Humans

Bird flu, though primarily an avian disease, poses a significant health risk to humans, especially those in close contact with infected birds or contaminated environments. Recognizing the symptoms early is crucial for timely medical intervention and preventing further spread.

Initial Symptoms

Bird flu symptoms in humans often begin similarly to other viral respiratory infections. These may include:

- **Fever:** A sudden onset of high fever, often exceeding 100.4°F (38°C).
- **Cough:** A persistent, dry cough is a common early symptom.
- **Sore Throat:** Irritation and pain in the throat can occur.
- **Muscle Aches:** Myalgia, or muscle pain, is frequently reported.
- **Fatigue:** Generalized weakness and lethargy often accompany the onset of bird flu.

Progressive Symptoms

As the infection progresses, more severe symptoms may develop, including:

- **Shortness of Breath:** Difficulty breathing, which can signal the onset of pneumonia.
- **Chest Pain:** Pain or discomfort in the chest due to lung involvement.

- **Diarrhea:** Gastrointestinal symptoms, such as diarrhea, are sometimes observed and may precede respiratory symptoms.
- **Conjunctivitis:** Red, swollen, and watery eyes are less common but significant indicators.

Prompt recognition of these symptoms, especially in individuals exposed to poultry or wild birds, can help reduce the severity of the disease and curb its spread.

4.2 Complications and Severe Cases

Bird flu can lead to severe health complications, particularly in individuals with pre-existing conditions or those who do not receive timely medical treatment. Understanding these complications is essential for effective management.

Severe Respiratory Issues

- **Pneumonia:** One of the most common and serious complications of bird flu. It results from the virus's ability to infect and inflame the lungs, causing fluid accumulation and impaired gas exchange.
- **Acute Respiratory Distress Syndrome (ARDS):** This life-threatening condition occurs when the lungs fail to provide adequate oxygen to the body due to severe inflammation and fluid leakage.

Multi-Organ Failure

In severe cases, the virus may spread beyond the respiratory system, leading to:

- **Kidney Failure:** Impaired kidney function due to systemic inflammation or direct viral damage.

- **Liver Dysfunction:** Elevated liver enzymes indicating hepatic involvement.
- **Heart Complications:** Myocarditis or heart failure in extreme cases.

Neurological Complications

Though rare, bird flu can affect the central nervous system, leading to:

- **Seizures:** Uncontrolled electrical disturbances in the brain.
- **Encephalitis:** Inflammation of the brain causing confusion, seizures, and coma.

High-Risk Groups

Certain individuals are more prone to severe complications:

- **Children and Elderly:** Weaker immune systems make these groups particularly vulnerable.
- **Immunocompromised Individuals:** People with weakened immune systems, such as those undergoing chemotherapy or living with HIV/AIDS.
- **Pregnant Women:** Physiological changes during pregnancy can increase susceptibility to complications.

Recognizing and managing these complications requires immediate medical attention, often in a hospital setting.

4.3 Differentiating Bird Flu from Seasonal Flu

Given the similarity in symptoms, distinguishing bird flu from seasonal influenza can be challenging. However, understanding the differences can guide clinicians and the general public in identifying the disease accurately.

Symptom Comparison

- **Fever:** Both bird flu and seasonal flu present with fever, but bird flu may cause higher and more prolonged fevers.

- **Respiratory Symptoms:** While both conditions cause cough and sore throat, bird flu often progresses more rapidly to severe respiratory issues like pneumonia.

- **Gastrointestinal Symptoms:** Diarrhea is more commonly associated with bird flu than seasonal flu.

- **Eye Infections:** Conjunctivitis is a notable symptom of bird flu but is rare in seasonal influenza.

Exposure History

A critical factor in differentiating the two diseases is the individual's exposure history:

- **Bird Flu:** Close contact with infected birds, contaminated surfaces, or visiting areas with known outbreaks increases the likelihood of bird flu.

- **Seasonal Flu:** Generally spreads via person-to-person contact and is more common during specific seasons.

Progression and Severity

Bird flu tends to progress more aggressively, often leading to severe complications such as multi-organ failure, which are less common in seasonal flu.

4.4 Diagnosis and Testing for Human Cases

Accurate diagnosis of bird flu in humans is essential for effective treatment and controlling the disease's spread. Diagnostic methods range from clinical evaluations to advanced laboratory tests.

Clinical Evaluation

Doctors typically begin with a thorough medical history and physical examination. Key factors include:

- **Symptom Assessment:** Detailed inquiry into respiratory and systemic symptoms.
- **Exposure History:** Questions about recent contact with poultry, wild birds, or affected areas.

Laboratory Tests

Laboratory testing provides definitive confirmation of bird flu. Common methods include:

- **RT-PCR (Reverse Transcription Polymerase Chain Reaction):** This is the gold standard for detecting avian influenza viruses. It identifies viral RNA with high specificity and sensitivity.
- **Viral Culture:** Although time-consuming, this method isolates and identifies the virus in a laboratory setting.
- **Serological Tests:** Detect antibodies produced in response to the virus, indicating a current or past infection.

Imaging Techniques

For patients presenting with severe respiratory symptoms, imaging studies like chest X-rays or CT scans can help assess lung involvement and identify complications like pneumonia.

Differential Diagnosis

It is crucial to rule out other diseases with similar presentations, such as:

- **Seasonal Flu:** Distinguished by its milder symptoms and seasonal patterns.
- **COVID-19:** Similar respiratory symptoms require testing for differentiation.
- **Bacterial Pneumonia:** May present similarly but responds to antibiotics, unlike viral infections.

Challenges in Diagnosis

Diagnosing bird flu in humans can be challenging due to:

- **Limited Access to Testing:** In resource-poor settings, advanced diagnostic tools may not be readily available.
- **Non-Specific Early Symptoms:** The similarity to common flu-like illnesses can delay diagnosis.
- **Rapid Progression:** The disease's aggressive nature requires quick and accurate identification.

Surveillance and Reporting

To enhance diagnostic accuracy and public health response, surveillance programs play a vital role. These include:

- **Monitoring Outbreaks:** Keeping track of avian influenza activity in bird populations.

- Global Collaboration: Sharing data across borders to predict and manage human cases.

- Public Awareness Campaigns: Educating at-risk populations about symptoms and when to seek medical attention.

In conclusion, understanding and recognizing bird flu symptoms in humans, differentiating it from other respiratory diseases, and implementing accurate diagnostic methods are critical steps in managing this zoonotic threat. Early detection not only improves patient outcomes but also helps prevent the disease's spread within communities.

Chapter 5: Transmission Dynamics

5.1 How Bird Flu Spreads Among Birds

Bird flu, or avian influenza, primarily spreads among birds through direct and indirect contact. Understanding these mechanisms is vital for controlling outbreaks in both wild and domestic bird populations.

Direct Contact

Infected birds shed the virus in their saliva, nasal secretions, and feces. Healthy birds contract the virus through:

- **Pecking and Preening:** Physical interactions such as pecking or preening feathers contaminated with the virus.

- **Fighting:** Aggressive encounters that lead to the exchange of fluids or close proximity.

Indirect Contact

The virus can also spread through contaminated surfaces and environments. Key factors include:

- **Contaminated Feed and Water:** Shared feeding and drinking areas act as hotspots for virus transmission.

- **Equipment and Clothing:** Farm tools, machinery, and clothing used in poultry farming can harbor and transfer the virus.

- **Aerosols:** Fine particles carrying the virus may spread in densely populated bird areas, such as commercial poultry farms.

Role of Vertical Transmission

In rare cases, the virus may pass from infected hens to their eggs, although this pathway is less common compared to horizontal transmission.

The dynamics of bird-to-bird transmission highlight the importance of stringent biosecurity measures in managing outbreaks.

5.2 Human Transmission Pathways

While bird flu primarily affects birds, human infections occur through specific pathways. Recognizing these transmission routes is crucial for preventing zoonotic spread.

Direct Contact with Birds

The most common route of transmission to humans involves close interaction with infected birds or their environments:

- **Handling Infected Birds:** Slaughtering, defeathering, and preparing poultry can expose individuals to the virus.

- **Exposure to Bird Droppings:** Contact with contaminated feces, often during cleaning or farming activities.

Environmental Contamination

The virus can survive in contaminated environments for extended periods, leading to:

- **Inhalation of Virus-Laden Particles:** Aerosolized particles from bird droppings or secretions may infect humans.

- **Touching Contaminated Surfaces:** Virus transfer occurs when individuals touch contaminated surfaces and then their face, mouth, or nose.

Consumption of Undercooked Poultry

Eating undercooked or raw poultry products from infected birds poses a risk, although proper cooking destroys the virus.

Limited Human-to-Human Transmission

While rare, human-to-human transmission has been reported in certain cases. These instances typically involve prolonged close contact, such as caregiving in households or healthcare settings.

Understanding these pathways informs public health strategies to reduce exposure risks for at-risk populations.

5.3 Zoonotic Risks and Public Health Concerns

The zoonotic nature of bird flu raises significant public health concerns, given its potential to cause pandemics.

Genetic Reassortment

The mixing of avian and human influenza viruses can create new strains capable of efficient human-to-human transmission. This reassortment poses a major threat to global health.

Occupational Risks

Certain professions are at higher risk of exposure, including:

- **Poultry Farmers:** Regular interaction with birds increases the likelihood of infection.
- **Slaughterhouse Workers:** Handling and processing infected birds heightens exposure.
- **Veterinarians:** Treating or examining birds during outbreaks.

Vulnerable Populations

Certain groups are more susceptible to severe outcomes if infected, such as:

- **Children and Elderly:** Weakened immune systems make these groups more vulnerable.

- **Individuals with Pre-Existing Conditions:** Chronic illnesses can exacerbate the severity of bird flu infections.

Potential for Global Spread

Bird flu's zoonotic potential, combined with global trade and travel, increases the risk of rapid international transmission. Migratory birds can carry the virus across borders, while infected poultry products may serve as vectors in the global supply chain.

Pandemic Preparedness

The possibility of a bird flu strain evolving into a highly transmissible form necessitates robust pandemic preparedness plans. These include:

- **Surveillance Programs:** Monitoring outbreaks in both avian and human populations.

- **Vaccination Development:** Proactive research into effective vaccines for emerging strains.

- **Public Education Campaigns:** Raising awareness about prevention and transmission mitigation.

5.4 Environmental Factors Affecting Spread

Environmental factors significantly influence the transmission dynamics of bird flu, shaping how and where the virus spreads.

Seasonal Patterns

Outbreaks often align with specific seasons due to environmental conditions:

- **Cold Weather:** The virus survives longer in cooler temperatures, increasing transmission rates during winter.
- **Migration Seasons:** Bird migrations in spring and autumn contribute to the virus's geographic spread.

Water Sources

Water bodies play a pivotal role in transmission:

- **Wild Birds and Waterfowl:** Lakes, rivers, and wetlands serve as gathering points for infected migratory birds, facilitating virus exchange.
- **Contaminated Drinking Water:** Shared water sources for domestic poultry can become reservoirs for the virus.

Farming Practices

The intensity and methods of poultry farming impact transmission dynamics:

- **High-Density Farming:** Crowded conditions in commercial poultry farms accelerate virus spread.
- **Poor Hygiene:** Lack of proper sanitation exacerbates contamination and transmission risks.

Urbanization and Habitat Disruption

Human activities that disrupt natural bird habitats can inadvertently increase virus spread:

- **Deforestation:** Forces wild birds into closer contact with domestic poultry.

- **Urban Expansion:** Creates interfaces where humans, wild birds, and poultry coexist, raising zoonotic risks.

Climate Change

Changes in climate patterns affect migratory behaviors and virus persistence:

- **Altered Migration Routes:** Shifts in migratory pathways can introduce the virus to new regions.

- **Extreme Weather Events:** Flooding and droughts can stress bird populations, making them more susceptible to infection.

By understanding and mitigating these environmental factors, policymakers and health organizations can develop targeted strategies to control bird flu outbreaks.

In summary, the transmission dynamics of bird flu involve complex interactions between biological, environmental, and human factors. Addressing these dynamics through effective surveillance, biosecurity, and public health initiatives is essential to mitigate the risks posed by this zoonotic disease.

Chapter 6: Prevention and Control in Birds

6.1 Best Practices in Biosecurity

Effective biosecurity measures are the cornerstone of preventing and controlling bird flu in avian populations. These practices aim to minimize the introduction and spread of the virus within and between bird populations.

Controlled Access to Farms

Restricting access to poultry farms is essential. Visitors and vehicles should be limited and disinfected before entering to reduce the risk of contamination. Establishing designated entry points and security perimeters ensures better control over potential disease vectors.

Personal Protective Equipment (PPE)

Farmworkers and visitors should use proper PPE, such as gloves, boots, and coveralls, which should be cleaned or disposed of after use. Changing into dedicated farm clothing helps reduce external contamination.

Sanitation Practices

Regular cleaning and disinfection of equipment, facilities, and vehicles help eliminate viral particles. Using proven disinfectants against avian influenza ensures the effectiveness of sanitation protocols. Water and feed sources should also be safeguarded from contamination by wild birds.

Isolation of New Birds

Introducing new birds to a flock requires a quarantine period to monitor for symptoms of disease. This minimizes the risk of introducing infected birds into a healthy population.

Wild Bird Deterrence

Creating barriers to prevent wild birds from mingling with domestic poultry is critical. This includes installing netting over outdoor enclosures and securing feed storage areas to avoid attracting wild birds.

Implementing these biosecurity measures creates a robust defense against the spread of bird flu within avian populations.

6.2 Vaccination in Poultry

Vaccination programs play a pivotal role in reducing the prevalence and impact of bird flu outbreaks in poultry.

Types of Vaccines

Two primary types of vaccines are used against avian influenza:

- **Inactivated Vaccines:** These contain killed virus particles and are administered through injection. They stimulate an immune response without causing disease.

- **Vector-Based Vaccines:** These use harmless viruses or bacteria to deliver avian influenza genes, prompting an immune response.

Implementation Strategies

Targeted Vaccination

In regions with high bird flu prevalence, targeted vaccination of at-risk poultry populations is critical. This includes:

- Birds in areas with frequent outbreaks.

- Commercial poultry farms near migratory bird pathways.

Mass Vaccination Campaigns

In severe outbreaks, mass vaccination campaigns are deployed to curb the spread. Vaccinating large flocks simultaneously reduces viral shedding and limits transmission.

Challenges in Vaccination

While vaccination is effective, it faces several hurdles:

- **Cost:** Widespread vaccination programs can be expensive, especially in low-resource areas.
- **Logistics:** Administering vaccines to large populations requires significant planning and manpower.
- **Viral Evolution:** Continuous monitoring is needed to ensure vaccine effectiveness against new strains.

A well-coordinated vaccination program, complemented by biosecurity, significantly enhances the control of bird flu in poultry.

6.3 Surveillance and Rapid Response

Timely detection and response are critical components of controlling bird flu outbreaks. Surveillance programs and rapid action plans are vital for minimizing the impact of the disease.

Surveillance Programs

Active Surveillance

Active surveillance involves regular testing of poultry and wild birds for avian influenza. This includes sampling blood, swabs from the respiratory tract, and feces to detect the virus.

Passive Surveillance

Farmers and poultry workers report suspected cases based on clinical symptoms, such as sudden deaths, reduced egg production, or respiratory distress. Passive surveillance complements active programs by flagging potential outbreaks.

Wild Bird Monitoring

Monitoring migratory bird populations provides early warning signals for potential disease spread to domestic poultry.

Rapid Response Measures

Quarantine and Culling

Once an outbreak is confirmed, affected areas are quarantined to prevent further spread. Infected flocks are humanely culled to eliminate the virus source.

Movement Restrictions

Restricting the movement of birds, feed, and equipment from affected areas limits the geographical spread of the virus.

Disinfection Campaigns

Thorough cleaning and disinfection of contaminated farms and equipment help eradicate residual viral particles.

Global Collaboration

International organizations, such as the World Organisation for Animal Health (WOAH) and the Food and Agriculture Organization (FAO), coordinate efforts to share data and resources, enabling better containment of outbreaks across borders.

6.4 Challenges in Eradicating Bird Flu

Despite advancements in biosecurity, vaccination, and surveillance, eradicating bird flu remains a formidable challenge.

Viral Mutation

Avian influenza viruses mutate rapidly, leading to new strains that may evade existing vaccines and detection methods. This genetic variability complicates long-term control efforts.

Migratory Bird Pathways

Migratory birds act as natural reservoirs for avian influenza. Their long-distance flights facilitate the spread of the virus across continents, introducing it to previously unaffected regions.

Economic Constraints

Resource limitations hinder the implementation of comprehensive prevention and control measures, particularly in developing countries. Costs associated with vaccination, surveillance, and outbreak management strain national budgets.

Cultural Practices

Traditional farming methods, live bird markets, and ceremonial practices involving birds increase the risk of virus transmission. Educating communities about safe practices is essential but often faces resistance due to deeply ingrained cultural norms.

Resistance to Culling

Culling infected and exposed birds is controversial, as it impacts farmers' livelihoods and food supply. Balancing disease control with economic and social considerations is a persistent challenge.

Gaps in Global Coordination

While international organizations play a significant role, discrepancies in surveillance and reporting standards between countries can delay response efforts. Harmonizing protocols and data sharing is essential for global eradication efforts.

In summary, controlling bird flu in avian populations requires a multifaceted approach, combining biosecurity, vaccination, surveillance, and international cooperation. Addressing the challenges head-on with innovative solutions and sustained commitment is crucial to mitigating the threat of this persistent disease.

Chapter 7: Prevention and Control in Humans

7.1 Personal Protective Measures

Personal protective measures are the first line of defense against bird flu in humans. These measures aim to reduce direct and indirect exposure to the avian influenza virus, particularly for individuals in high-risk occupations or living near outbreaks.

Hygiene Practices

Maintaining proper hygiene is one of the most effective ways to prevent infection:

- **Handwashing:** Regularly washing hands with soap and water, especially after handling birds or poultry products, is crucial. Alcohol-based hand sanitizers can be used when soap and water are unavailable.

- **Avoid Touching the Face:** Minimizing contact with the eyes, nose, and mouth can prevent the virus from entering the body.

- **Use of Masks:** Wearing surgical or N95 masks in high-risk areas helps block respiratory droplets that may carry the virus.

Protective Gear

For those working in poultry farming or avian research, wearing protective gear is essential:

- **Gloves:** Disposable gloves prevent direct contact with infected birds or contaminated surfaces.

- **Boot Covers:** Protecting footwear reduces the risk of spreading the virus to other locations.

- **Goggles:** Eye protection is necessary to prevent exposure to respiratory droplets or splashes.

Avoiding Risky Behaviors

Reducing exposure to potentially infected birds can lower the risk of infection:

- **Avoid Handling Dead or Sick Birds:** Reporting such birds to local authorities rather than handling them ensures proper containment.

- **Cooking Poultry Thoroughly:** Consuming well-cooked poultry and eggs eliminates the risk of infection through contaminated food.

By adopting these personal protective measures, individuals can significantly reduce their likelihood of contracting bird flu.

7.2 Vaccines and Antiviral Medications

Vaccination and antiviral medications play critical roles in controlling bird flu outbreaks and preventing severe cases in humans.

Current Vaccine Options

Although vaccines for human use against bird flu are limited, research is ongoing to develop effective options:

- **H5N1 Vaccines:** Several vaccines targeting the H5N1 strain have been approved for use in specific circumstances. These are often reserved for individuals at high risk, such as poultry workers and healthcare providers.

- **Universal Flu Vaccines:** Efforts are underway to create broad-spectrum vaccines that protect against multiple influenza subtypes, including avian strains.

Challenges in Vaccine Development

Developing effective vaccines against bird flu is complicated due to:

- **Viral Mutations:** Rapid genetic changes in avian influenza viruses can render existing vaccines less effective.

- **Global Availability:** Limited production and distribution capacity hinder widespread access, especially in low-income regions.

Antiviral Medications

Antiviral drugs can be used to treat or prevent bird flu infections in humans:

- **Oseltamivir (Tamiflu):** This drug reduces the severity and duration of symptoms if administered within 48 hours of onset.

- **Zanamivir (Relenza):** Another option for treating bird flu, administered via inhalation.

- **Prophylactic Use:** Antivirals can be prescribed preventively to individuals exposed to the virus, such as healthcare workers.

Resistance Concerns

The overuse of antivirals has led to concerns about resistance:

- **Emerging Resistant Strains:** Some bird flu strains have shown reduced susceptibility to common antivirals, emphasizing the need for judicious use and continued research.

Vaccines and antiviral medications remain vital tools in reducing the impact of bird flu on human populations, but their effectiveness depends on timely administration and ongoing innovation.

7.3 Public Awareness Campaigns

Public awareness campaigns are crucial in educating communities about bird flu risks and prevention strategies. These campaigns aim to empower individuals with knowledge to protect themselves and their families.

Objectives of Awareness Campaigns

Key goals include:

- **Promoting Hygiene Practices:** Teaching proper handwashing techniques and safe poultry handling.

- **Encouraging Early Reporting:** Emphasizing the importance of reporting sick or dead birds to authorities.

- **Dispelling Myths:** Correcting misinformation about bird flu transmission to reduce panic and stigma.

Effective Communication Strategies

Reaching diverse populations requires tailored approaches:

- **Multimedia Outreach:** Using television, radio, social media, and print materials ensures widespread dissemination of information.

- **Community Engagement:** Partnering with local leaders and organizations fosters trust and facilitates information sharing.

- **Visual Aids:** Infographics and videos simplify complex information for broader understanding.

Target Audiences

Awareness campaigns should focus on groups most at risk:

- **Rural Communities:** Often in close contact with poultry, these populations benefit from education on biosecurity measures.

- **Healthcare Workers:** Training programs help professionals recognize symptoms and follow safety protocols.

- **Travelers:** Informing travelers about outbreak regions and preventive measures reduces the risk of international spread.

Measuring Success

The effectiveness of awareness campaigns can be evaluated through:

- **Increased Reporting Rates:** A rise in reported bird flu cases indicates improved community vigilance.

- **Behavioral Changes:** Observing higher compliance with hygiene practices and protective measures.

- **Reduction in Outbreaks:** Fewer human cases reflect the success of preventive strategies.

Public awareness campaigns are an indispensable part of comprehensive bird flu prevention efforts, bridging the gap between scientific knowledge and community action.

7.4 Role of Global Health Organizations

Global health organizations play a pivotal role in coordinating international efforts to prevent and control bird flu outbreaks.

Surveillance and Monitoring

Organizations like the World Health Organization (WHO) and the Food and Agriculture Organization (FAO) oversee global surveillance programs:

- **Tracking Outbreaks:** Monitoring avian influenza cases in birds and humans helps identify emerging hotspots.

- **Data Sharing:** Collaborating with national governments ensures timely dissemination of information.

- **Early Warning Systems:** Predicting potential outbreaks allows for preemptive action.

Policy Development

Global health bodies provide guidelines to standardize prevention and control measures:

- **Biosecurity Protocols:** Recommendations for poultry farms to minimize infection risks.

- **Travel Advisories:** Informing travelers about affected regions and preventive steps.
- **Vaccination Policies:** Advising on vaccine stockpiling and distribution priorities.

Resource Mobilization

During outbreaks, international organizations mobilize resources to support affected regions:

- **Funding:** Providing financial aid for outbreak response and research.
- **Technical Expertise:** Deploying experts to assist with containment and treatment efforts.
- **Supply Distribution:** Ensuring access to vaccines, antivirals, and protective equipment.

Research and Innovation

Global organizations facilitate research into bird flu prevention and treatment:

- **Collaborative Studies:** Partnering with academic institutions and pharmaceutical companies to develop new interventions.
- **Emerging Threats:** Identifying and addressing novel strains of avian influenza.

Challenges in Global Coordination

Despite significant progress, several challenges remain:

- **Political Barriers:** Differences in national policies and priorities can hinder collaboration.
- **Resource Inequities:** Disparities in funding and infrastructure limit response capabilities in low-income countries.
- **Public Compliance:** Achieving global adherence to preventive measures requires sustained educational efforts.

Global health organizations serve as the backbone of international bird flu prevention and control, ensuring a coordinated, science-based approach to mitigate risks.

By integrating personal protective measures, advancements in vaccines and antivirals, effective public awareness campaigns, and the coordinated efforts of global health organizations, humanity can significantly reduce the impact of bird flu. Comprehensive prevention and control strategies not only protect individuals but also strengthen public health systems against future outbreaks.

Chapter 8: Impact on Global Health and Economy

8.1 Economic Costs of Bird Flu Outbreaks

Bird flu outbreaks have profound economic repercussions at local, national, and global levels. The economic costs are felt across multiple sectors, including agriculture, healthcare, and international trade.

Agricultural Losses

The poultry industry is among the hardest hit by bird flu outbreaks. Key impacts include:

- **Mass Culling of Poultry:** To contain the spread of the virus, millions of birds are often culled. This results in direct financial losses for farmers and poultry companies.

- **Disruption of Production:** Outbreaks lead to decreased poultry production, affecting both meat and egg supplies.

- **Increased Biosecurity Costs:** Farmers and producers must invest in enhanced biosecurity measures, such as sanitization, protective equipment, and monitoring systems, which add to operational costs.

International Trade Restrictions

Many countries impose bans on poultry imports from regions affected by bird flu, leading to:

- **Loss of Export Revenue:** Poultry-exporting countries suffer significant financial losses when trade restrictions are implemented.

- **Market Instability:** The sudden drop in supply and demand disrupts global poultry markets, causing price volatility.

Healthcare Expenditures

The emergence of bird flu in humans incurs substantial costs for healthcare systems:

- **Medical Treatment:** Treating infected individuals and managing complications can strain healthcare budgets.

- **Surveillance and Research:** Governments and organizations allocate resources to monitor outbreaks, study the virus, and develop vaccines and treatments.

Indirect Economic Impacts

The economic effects extend beyond agriculture and healthcare:

- **Loss of Livelihoods:** Farmers, workers, and businesses involved in the poultry supply chain face job losses and financial instability.

- **Tourism Declines:** Outbreaks can deter tourists from visiting affected regions, further affecting local economies.

Addressing these economic challenges requires coordinated efforts to mitigate outbreaks and support affected industries and communities.

8.2 Effects on Food Security

Bird flu outbreaks significantly impact food security, particularly in regions where poultry is a staple protein source.

Reduced Poultry Availability

The mass culling of infected birds and trade restrictions lead to:

- **Supply Shortages:** Decreased availability of poultry meat and eggs, causing price hikes.

- **Protein Deficiency:** Communities reliant on poultry as their primary protein source face nutritional challenges.

Price Increases

The disruption of poultry production and trade often results in higher prices for consumers:

- **Affordability Issues:** Low-income households struggle to afford alternative protein sources, exacerbating food insecurity.

- **Market Inequality:** Wealthier nations can offset shortages with imports, while poorer countries face prolonged supply disruptions.

Impact on Small-Scale Farmers

Small-scale poultry farmers are particularly vulnerable to the effects of bird flu outbreaks:

- **Loss of Livelihood:** Small farmers often lack the resources to recover from culling and production losses.

- **Food Access:** These farmers may also depend on their poultry for personal consumption, further affecting their food security.

Addressing food security concerns during bird flu outbreaks requires targeted interventions, such as financial aid for farmers, alternative protein programs, and strategies to stabilize markets.

8.3 Impact on Wildlife Conservation

Bird flu poses a significant threat to wildlife conservation efforts, particularly for avian species.

Threats to Wild Bird Populations

Wild birds are natural hosts for avian influenza viruses, and outbreaks can devastate populations:

- **Mass Mortality Events:** Certain strains of bird flu cause high mortality rates among wild birds, including endangered species.

- **Disruption of Ecosystems:** The loss of key bird species can disrupt ecological balance, affecting predator-prey relationships and ecosystem services.

Risks to Migratory Birds

Migratory birds play a critical role in spreading bird flu across regions:

- **Infection During Migration:** Infected birds can spread the virus to new locations, affecting local wildlife.

- **Impact on Conservation Programs:** Efforts to protect migratory routes and habitats are complicated by the need to manage and monitor outbreaks.

Challenges for Conservation Organizations

Organizations dedicated to wildlife conservation face unique challenges during bird flu outbreaks:

- **Resource Diversion:** Limited resources must be allocated to monitoring and controlling the virus, detracting from other conservation initiatives.

- **Balancing Conservation and Public Health:** Managing bird flu in wild populations often involves difficult decisions, such as culling or habitat restrictions, which can conflict with conservation goals.

Conservation strategies must consider the dual objectives of protecting avian species and preventing virus transmission to domestic birds and humans.

8.4 Pandemic Preparedness and Lessons Learned

Bird flu outbreaks serve as critical learning opportunities for improving global pandemic preparedness.

Strengthening Surveillance Systems

Effective surveillance is essential for early detection and containment of outbreaks:

- **Integrated Monitoring:** Collaboration between agricultural, veterinary, and public health sectors enhances surveillance capabilities.

- **Global Networks:** International initiatives, such as the Global Influenza Surveillance and Response System (GISRS), facilitate data sharing and coordination.

Vaccine Development

Advances in vaccine technology play a crucial role in pandemic preparedness:

- **Targeted Vaccines:** Developing vaccines for both poultry and humans helps prevent and control outbreaks.

- **Stockpiling and Distribution:** Ensuring adequate vaccine supplies and equitable distribution reduces response times during outbreaks.

Public Awareness and Education

Educating communities about bird flu risks and prevention measures is vital:

- **Behavioral Change:** Promoting safe handling practices and hygiene reduces exposure risks.

- **Combatting Misinformation:** Public awareness campaigns address myths and misconceptions about the virus.

Policy and International Collaboration

Governments and organizations must adopt policies that prioritize pandemic preparedness:

- **Emergency Response Plans:** Establishing clear protocols for outbreak containment minimizes delays.

- **International Cooperation:** Coordinated efforts between countries enhance resource sharing and outbreak management.

Lessons for Future Pandemics

Bird flu outbreaks offer valuable insights into managing future pandemics:

- **Proactive Approaches:** Investing in prevention and preparedness reduces the likelihood of widespread outbreaks.

- **One Health Approach:** Recognizing the interconnectedness of human, animal, and environmental health fosters holistic strategies.

In summary, bird flu's impact on global health and the economy underscores the need for comprehensive prevention and control measures. By learning from past outbreaks, the global community can enhance its preparedness for future challenges, ensuring a more resilient and coordinated response.

Chapter 9: Future Outlook

9.1 Research Trends in Avian Influenza

Research on avian influenza (AI) continues to evolve, reflecting the urgency of understanding and mitigating this persistent threat. Scientists and organizations worldwide are investing in innovative methods to track, study, and combat the virus.

Advances in Virology

- **Genomic Sequencing:** Rapid sequencing technologies enable researchers to monitor genetic changes in avian influenza strains, facilitating early detection of potential pandemic variants.

- **Host-Virus Interactions:** Studies on how the virus interacts with avian and human hosts are uncovering mechanisms that drive transmission and pathogenicity.

- **Mutational Studies:** Research focuses on mutations that allow the virus to cross species barriers and adapt to human hosts.

Predictive Modeling

- **Epidemiological Models:** Advanced models predict outbreak dynamics based on environmental, migratory, and economic data.

- **Machine Learning:** AI tools analyze large datasets to identify patterns and forecast potential hotspots for outbreaks.

Integrated Surveillance Systems

- **Real-Time Monitoring:** Satellite and drone technologies are being used to track migratory bird populations and potential AI reservoirs.

- **Interdisciplinary Collaboration:** Combining insights from ecology, virology, and public health enhances surveillance efficacy.

Challenges in Research

Despite advancements, challenges remain:

- **Funding Gaps:** Sustained research funding is essential for long-term progress.

- **Global Collaboration:** Coordinating international research efforts is often hindered by political and logistical barriers.

- **Ethical Considerations:** Studies involving live animals or potentially pandemic strains must navigate complex ethical and safety concerns.

9.2 Genetic Changes and Vaccine Development

The genetic plasticity of avian influenza viruses presents both challenges and opportunities in vaccine development. Continuous monitoring and adaptation are necessary to stay ahead of emerging strains.

Genetic Evolution of AI Viruses

- **Antigenic Drift and Shift:** Minor genetic changes (drift) and major reassortments (shift) contribute to the virus's ability to evade immunity.

- **Reassortment Events:** Co-infection in animals can lead to new hybrid strains, increasing pandemic potential.
- **Zoonotic Transmission:** Genetic adaptations that enhance binding to human receptors are a key focus of research.

Innovations in Vaccine Technology

- **Universal Vaccines:** Researchers aim to develop vaccines targeting conserved viral elements to provide broad-spectrum protection.
- **mRNA Vaccines:** Building on the success of COVID-19 vaccines, mRNA technology offers rapid and scalable solutions for AI.
- **Recombinant Vaccines:** Using viral vectors or proteins, these vaccines are safer and more effective for poultry and humans.

Challenges in Vaccine Deployment

- **Production and Distribution:** Meeting global demand requires robust manufacturing and equitable distribution systems.
- **Vaccine Hesitancy:** Public mistrust and misinformation can hinder vaccination campaigns.
- **Strain-Specific Efficacy:** Rapid genetic changes necessitate frequent updates to vaccine formulations.

9.3 Cross-Sector Collaborations in Disease Management

Effective management of avian influenza requires collaboration across sectors, reflecting the interconnected nature of human, animal, and environmental health.

One Health Approach

- **Holistic Framework:** This approach integrates human, animal, and environmental health to address zoonotic diseases like AI.

- **Key Stakeholders:** Governments, non-governmental organizations (NGOs), academic institutions, and industries must work together.

Public-Private Partnerships (PPPs)

- **Resource Mobilization:** Collaborations with the private sector can provide funding, expertise, and technological innovations.

- **Example Initiatives:** Partnerships for vaccine production, diagnostic tool development, and outbreak response strategies have proven effective.

Regional and Global Cooperation

- **International Organizations:** Entities like the World Health Organization (WHO), Food and Agriculture Organization (FAO), and World Organisation for Animal Health (WOAH) play critical roles in coordinating global efforts.

- **Regional Networks:** Platforms like the Association of Southeast Asian Nations (ASEAN) promote localized strategies for outbreak control.

Community Engagement

- **Grassroots Involvement:** Engaging local communities ensures compliance with prevention measures and enhances outbreak reporting.

- **Education Campaigns:** Training programs for farmers, healthcare workers, and public health officials bridge knowledge gaps.

9.4 Challenges in Predicting Future Outbreaks

The unpredictable nature of avian influenza outbreaks underscores the need for continual vigilance and adaptive strategies.

Complexity of Disease Ecology

- **Reservoir Dynamics:** Wild birds and other animal reservoirs contribute to the virus's persistence and spread.

- **Environmental Factors:** Climate change, habitat destruction, and urbanization alter the ecological balance, influencing outbreak patterns.

Human Factors

- **Global Travel and Trade:** Increased connectivity accelerates the spread of AI across borders.

- **Agricultural Practices:** Intensive poultry farming and live bird markets amplify transmission risks.

- **Healthcare Inequities:** Limited resources in low-income countries hinder outbreak detection and response.

Technological Limitations

- **Data Gaps:** Incomplete or inconsistent data impairs predictive modeling accuracy.

- **Diagnostic Challenges:** Early-stage infections are often asymptomatic, complicating timely detection.

- **Resource Constraints:** Underfunded health systems struggle to implement advanced monitoring tools.

Strategies for Improvement

To address these challenges, the global community must:

- **Invest in Research and Development:** Prioritize funding for innovative surveillance and diagnostic technologies.

- **Enhance Data Sharing:** Foster transparent and real-time information exchange among countries and organizations.

- **Strengthen Global Health Infrastructure:** Build capacity for rapid response and containment of outbreaks.

- **Promote Sustainable Practices:** Encourage environmentally friendly agricultural methods to reduce zoonotic spillovers.

Conclusion

The future of avian influenza management hinges on a proactive and collaborative approach. By advancing research, refining vaccines, fostering cross-sector partnerships, and addressing prediction challenges, the global community can mitigate the threat posed by AI. Lessons learned from past outbreaks and current innovations provide a foundation for resilience, ensuring preparedness for whatever challenges lie ahead.

Chapter 10: Conclusion and Key Takeaways

10.1 Summary of Key Points

Avian influenza (AI) represents a persistent global threat with profound implications for human and animal health, economic stability, and food security. The preceding chapters have explored the multifaceted challenges and strategies associated with AI management, emphasizing the importance of a unified, interdisciplinary approach. Key takeaways include:

- **Surveillance and Early Detection:** Real-time monitoring and predictive modeling are vital for identifying and mitigating outbreaks before they escalate.

- **Genetic Research:** Understanding viral mutations and their impact on transmission dynamics is crucial for effective vaccine development and therapeutic interventions.

- **Vaccine Innovations:** Advances in mRNA and universal vaccine technologies hold promise for addressing the evolving nature of AI.

- **One Health Approach:** Integrating human, animal, and environmental health perspectives fosters holistic and sustainable solutions.

- **Collaboration and Community Engagement:** Regional, national, and international partnerships, alongside grassroots efforts, amplify the effectiveness of prevention and control measures.

10.2 Importance of Continued Vigilance

While significant progress has been made, the unpredictable nature of AI necessitates unwavering vigilance. Several factors underscore this need:

- **Rapid Viral Evolution:** The high mutation rate of AI viruses increases the likelihood of new strains emerging, with potential pandemic implications.

- **Environmental and Human Factors:** Climate change, habitat destruction, and intensive farming practices create conditions conducive to viral spillovers.
- **Global Connectivity:** Modern travel and trade networks accelerate the spread of infectious diseases, amplifying their impact.

Continued investment in surveillance systems, research, and public health infrastructure is non-negotiable. Policymakers, researchers, and communities must remain alert to emerging trends and adapt strategies accordingly.

10.3 Role of Individuals in Prevention

Individual actions are integral to controlling the spread of avian influenza. Empowering people with knowledge and resources enhances collective resilience. Key roles individuals can play include:

Practicing Biosecurity Measures

- **Hygiene:** Frequent handwashing and proper sanitation reduce the risk of infection.
- **Safe Handling of Poultry:** Avoiding contact with sick or dead birds and ensuring proper cooking of poultry products minimizes exposure.

Supporting Sustainable Practices

- **Consumer Choices:** Opting for poultry from farms adhering to biosecurity standards encourages industry-wide improvements.
- **Advocacy:** Advocating for sustainable farming and conservation efforts helps mitigate environmental factors contributing to AI outbreaks.

Staying Informed

- **Public Health Alerts:** Keeping abreast of updates from trusted sources like the World Health Organization (WHO) and Centers for Disease Control and Prevention (CDC) ensures timely action during outbreaks.

- **Community Engagement:** Participating in local awareness campaigns and training programs strengthens collective preparedness.

10.4 Building Resilience Against Emerging Diseases

The fight against avian influenza provides valuable lessons for managing other emerging infectious diseases. Building resilience involves:

Strengthening Public Health Systems

- **Infrastructure Development:** Equipping healthcare facilities with the tools and resources needed to respond to outbreaks promptly.

- **Training and Capacity Building:** Enhancing the skills of healthcare professionals and first responders.

Promoting Research and Innovation

- **Interdisciplinary Collaboration:** Encouraging partnerships between virologists, epidemiologists, and data scientists to accelerate discoveries.

- **Investing in Technology:** Leveraging AI and machine learning for predictive modeling, diagnostics, and vaccine development.

Enhancing Global Solidarity

- **Equitable Resource Distribution:** Ensuring that low- and middle-income countries have access to vaccines, treatments, and diagnostic tools.

- **Transparent Communication:** Sharing data and best practices fosters trust and efficiency in managing global health crises.

Final Thoughts

Avian influenza serves as a stark reminder of the interconnectedness of human, animal, and environmental health. Addressing this complex challenge requires a comprehensive, inclusive, and forward-thinking approach. By integrating technological advancements, fostering collaboration, and empowering individuals, society can mitigate the threat of AI and other emerging diseases. The journey toward resilience is ongoing, but every step taken today strengthens the foundation for a healthier and safer tomorrow.

Further Resources

For those interested in deepening their understanding of avian influenza and related topics, the following resources are recommended:

Books and Articles

1. "The Coming Plague" by Laurie Garrett
   - An in-depth exploration of emerging infectious diseases and their global impact.

2. "Spillover: Animal Infections and the Next Human Pandemic" by David Quammen
   - A compelling narrative on zoonotic diseases, including avian influenza.

3. Scientific Journals:
   - *Journal of Virology*
   - *Emerging Infectious Diseases*
   - *The Lancet Infectious Diseases*

Organizations and Websites

1. World Health Organization (WHO):
   - Provides comprehensive updates and guidelines on avian influenza.
   - Website: www.who.int

2. Centers for Disease Control and Prevention (CDC):
   - Offers resources for prevention, surveillance, and outbreak response.
   - Website: www.cdc.gov

3. Food and Agriculture Organization (FAO):
   - Focuses on the intersection of animal health and food security.
   - Website: www.fao.org

4. **World Organisation for Animal Health (WOAH):**
    - Dedicated to animal disease monitoring and management.
    - Website: www.woah.org

Online Learning Platforms

1. **Coursera and edX:**
    - Courses on infectious disease management, virology, and global health.

2. **FAO eLearning:**
    - Modules on animal health and biosecurity measures.

3. **WHO Academy:**
    - Offers training resources for healthcare professionals and public health workers.

Community and Advocacy Groups

1. **Local Public Health Departments:**
    - Provide workshops and training on disease prevention and biosecurity.

2. **One Health Initiative:**
    - Promotes interdisciplinary approaches to health challenges.

3. **Non-Governmental Organizations (NGOs):**
    - Entities like Doctors Without Borders support outbreak response efforts globally.

By utilizing these resources, individuals and organizations can play a pivotal role in enhancing preparedness and resilience against avian influenza and other emerging health threats.

References

1. Alexander, D. J. (2007). *An overview of the epidemiology of avian influenza.* Vaccine, 25(30), 5637-5644. https://doi.org/10.1016/j.vaccine.2006.10.051

2. Capua, I., & Alexander, D. J. (2009). *Avian influenza and human health.* Acta Tropica, 110(1), 8-12. https://doi.org/10.1016/j.actatropica.2008.09.010

3. Centers for Disease Control and Prevention (CDC). (2024). *Avian Influenza: Information and updates.* Retrieved from www.cdc.gov

4. Food and Agriculture Organization of the United Nations (FAO). (2023). *Avian influenza prevention and control: Guidelines and strategies.* Retrieved from www.fao.org

5. Garamszegi, L. Z., & Møller, A. P. (2007). *Prevalence of avian influenza and host ecology.* Proceedings of the Royal Society B, 274(1609), 2003-2012. https://doi.org/10.1098/rspb.2007.0053

6. Quammen, D. (2012). *Spillover: Animal infections and the next human pandemic.* W.W. Norton & Company.

7. Swayne, D. E., & Halvorson, D. A. (2003). *Influenza.* In Y. M. Saif (Ed.), *Diseases of poultry* (11th ed., pp. 135-160). Iowa State Press.

8. World Health Organization (WHO). (2024). *Avian influenza fact sheet.* Retrieved from www.who.int

9. Webster, R. G., & Govorkova, E. A. (2006). *H5N1 influenza: Continuing evolution and spread.* New England Journal of Medicine, 355(21), 2174-2177. https://doi.org/10.1056/NEJMp068249

10. World Organisation for Animal Health (WOAH). (2023). *Avian influenza (bird flu): Current status and measures.* Retrieved from www.woah.org

Author's Note

Writing this book, **"Understanding Bird Flu: Causes, Symptoms, and Transmission,"** has been a journey of discovery and a deep dive into one of the most pressing global health challenges of our time. The intricacies of avian influenza—its causes, transmission, and far-reaching impacts—serve as a testament to the delicate balance of our interconnected world. This book was born out of a desire to demystify the science behind avian influenza and provide actionable insights for researchers, policymakers, healthcare professionals, and the general public.

The significance of avian influenza cannot be overstated. Its effects ripple through public health systems, economies, ecosystems, and individual lives. My hope is that this book not only informs but also inspires proactive measures to mitigate its impact.

I extend my deepest gratitude to the scientists, public health professionals, and global organizations whose tireless work forms the backbone of our understanding of avian influenza. Their research and innovations are guiding lights in the fight against infectious diseases.

I also thank you, the reader, for your interest in this topic. By seeking knowledge, you are contributing to a world that is better prepared to address the challenges posed by avian influenza and other emerging diseases. I encourage you to use this information as a starting point to explore, advocate, and act in ways that promote global health and resilience.

If this book sparks even a single conversation about disease prevention, public health, or global collaboration, then it has achieved its purpose. Together, through awareness and action, we can build a healthier, safer, and more informed world.

Thank you for joining me on this journey.

Sincerely,

Oluchi Ike

www.ingramcontent.com/pod-product-compliance
Lightning Source LLC
Chambersburg PA
CBHW062117220526
45471CB00010B/3773